Energy Sector Standard of the People's Republic of China

NB/T 10394-2020

# Specification for photovoltaic power generation system performance

# 光伏发电系统效能规范

*(English Translation)*

China Water & Power Press
中国水利水电出版社
Beijing 2024

**图书在版编目（CIP）数据**
光伏发电系统效能规范 : NB/T 10394-2020 = Specification for photovoltaic power generation system performance （NB/T 10394-2020） : 英文 / 国家能源局发布. -- 北京 : 中国水利水电出版社, 2024. 8. -- ISBN 978-7-5226-2710-6
Ⅰ. TM615-65
中国国家版本馆CIP数据核字第20248DJ838号

Energy Sector Standard of the People's Republic of China
中华人民共和国能源行业标准

**Specification for photovoltaic power generation system performance**
**光伏发电系统效能规范**

**NB/T 10394-2020**

( English Translation )

Issued by National Energy Administration of the People's Republic of China
国家能源局　发布
Translation organized by China Renewable Energy Engineering Institute
水电水利规划设计总院　组织翻译
Published by China Water & Power Press
中国水利水电出版社　出版发行
Tel: (+ 86 10) 68545888　68545874
sales@mwr.gov.cn
Account name: China Water & Power Press
Address: No.1, Yuyuantan Nanlu, Haidian District, Beijing 100038, China
http: //www.waterpub.com.cn
中国水利水电出版社微机排版中心　排版
北京中献拓方科技发展有限公司　印刷
210mm×297mm　16 开本　1.75 印张　71 千字
2024 年 8 月第 1 版　2024 年 8 月第 1 次印刷
**Price（定价）: ¥ 285.00**

# About English Translation

This English version is one of China's energy sector standard series in English. Its translation was organized by China Renewable Energy Engineering Institute authorized by National Energy Administration of the People's Republic of China in compliance with relevant procedures and stipulations. This English version was issued by National Energy Administration of the People's Republic of China in Announcement [2023] No. 8 dated December 28, 2023.

This version was translated from the Chinese Standard NB/T 10394-2020, *Specification for photovoltaic power generation system performance*, published by China Water & Power Press. The copyright is reserved by National Energy Administration of the People's Republic of China. In the event of any discrepancy in the implementation, the Chinese version shall prevail.

Many thanks go to the staff from the relevant standard development organizations and those who have provided generous assistance in the translation and review process.

For further improvement of the English version, any comments and suggestions are welcome and should be addressed to:

China Renewable Energy Engineering Institute
No. 2 Beixiaojie, Liupukang, Xicheng District, Beijing 100120, China
Website: www.creei.cn

Translating organizations:

POWERCHINA Northwest Engineering Corporation Limited

China Renewable Energy Engineering Institute

Translating staff:

| | | | |
|---|---|---|---|
| YAN Cunku | XIAO Bin | LI Cheng | QIN Xiao |
| LYU Song | LIU Xiaoxin | DU Wei | AN Zaizhan |
| LIU Xiangyang | | | |

Review panel members:

| | |
|---|---|
| GUO Jie | POWERCHINA Beijing Engineering Corporation Limited |
| QIE Chunsheng | Senior English Translator |
| YAN Wenjun | Army Academy of Armored Forces, PLA |
| ZHANG Ming | Tsinghua University |
| QI Wen | POWERCHINA Beijing Engineering Corporation Limited |
| CHEN Lei | POWERCHINA Zhongnan Engineering Corporation Limited |
| LIANG Hongli | Shanghai Investigation, Design & Research Institute Co., Ltd. |
| YUE Lei | China Renewable Energy Engineering Institute |

National Energy Administration of the People's Republic of China

# 翻译出版说明

本译本为国家能源局委托水电水利规划设计总院按照有关程序和规定，统一组织翻译的能源行业标准英文版系列译本之一。2023 年 12 月 28 日，国家能源局以 2023 年第 8 号公告予以公布。

本译本是根据中国水利水电出版社出版的《光伏发电系统效能规范》NB/T 10394—2020 翻译的，著作权归国家能源局所有。在使用过程中，如出现异议，以中文版为准。

本译本在翻译和审核过程中，本标准编制单位及编制组有关成员给予了积极协助。

为不断提高本译本的质量，欢迎使用者提出意见和建议，并反馈给水电水利规划设计总院。

地址：北京市西城区六铺炕北小街 2 号
邮编：100120
网址：www.creei.cn

本译本翻译单位：中国电建集团西北勘测设计研究院有限公司
水电水利规划设计总院

本译本翻译人员：严存库　肖　斌　李　诚　秦　潇
吕　嵩　刘孝鑫　杜　为　安再展
刘向阳

本译本审核人员：

郭　洁　中国电建集团北京勘测设计研究院有限公司

郄春生　英语高级翻译

闫文军　中国人民解放军陆军装甲兵学院

张　明　清华大学

齐　文　中国电建集团北京勘测设计研究院有限公司

陈　蕾　中国电建集团中南勘测设计研究院有限公司

梁洪丽　上海勘测设计研究院有限公司

岳　蕾　水电水利规划设计总院

国家能源局

# Contents

# Foreword

This document is drafted in accordance with the rules given in the GB/T 1.1-2009 *Directives for standardization—Part 1: Structure and drafting of standards*.

Attention is drawn to the possibility that some of the elements of this document may be the subject of patent rights. Issuing body shall not be held responsible for identifying any or all such patent rights.

National Energy Administration of the People's Republic of China is in charge of the administration of this document. China Renewable Energy Engineering Institute has proposed this document and is responsible for its routine management. China Renewable Energy Engineering Institute is responsible for the explanation of specific technical contents. Comments and suggestions in the implementation of this document should be addressed to:

China Renewable Energy Engineering Institute
No. 2 Beixiaojie, Liupukang, Xicheng District, Beijing 100120, China

Drafting organizations:

China Renewable Energy Engineering Institute
POWERCHINA Northwest Engineering Corporation Limited
POWERCHINA Shanghai Electric Power Engineering Co., Ltd.
China Photovoltaic Industry Association
China Three Gorges Renewables (Group) Co., Ltd.
China Huaneng Group Co., Ltd.
LONGi Green Energy Technology Co., Ltd.
TONGWEI Group Co., Ltd.
SUNGROW Power Supply Co., Ltd.
HUAWEI Technologies Co., Ltd.
SINENG Electric Co., Ltd.

Chief drafting staff:

| | | | |
|---|---|---|---|
| QIN Xiao | LYU Song | CHEN Gang | ZHANG Bo |
| WANG Jixue | XIAO Bin | WANG Haoyi | JIN Yanmei |
| FU Zhengning | ZHOU Zhi | HUI Xing | DU Wei |
| WANG Shuo | CUI Yun | LIU Jianping | NI Xin |
| CUI Yongfeng | LIU Xiaoxin | LI Cheng | TIAN Lisha |
| DONG Feifei | ZHAO Wei | LI Dongxia | REN Jiachen |
| LIU Yinglian | LU Guocheng | NIU Zhiyuan | LIU Songmin |
| ZHANG Ling | ZHAO Wei | SHU Zhenhuan | WANG Li |

# Introduction

The photovoltaic power generation system performance is a system of indicators for comprehensive assessment on the techno-economy of the photovoltaic power generation system. It is an important reference for investment analysis, design optimization, quality management, and operation and maintenance (O&M) of the photovoltaic power generation system, and an important basis for acceptance and post-assessment of the system.

In order to standardize and guide the assessment on photovoltaic power generation system performance, standardize the assessment indicators, calculation methods and performance assessment process, and establish the performance assessment system this document has been developed according to the requirements of Document GNZTKJ [2018] No. 191 "Notice on Releasing the Supplementary Plan for the Development and Revision of Energy Sector Standards (the Second Batch) in 2018" issued by General Department of National Energy Administration of the People's Republic of China.

# Specification for photovoltaic power generation system performance

## 1 Scope

This document specifies the basic assessment requirements, assessment methods and assessment process of photovoltaic power generation system performance.

This document is applicable to the assessment on techno-economy of photovoltaic power generation system.

## 2 Normative references

The following referenced documents are indispensable for the application of this document. For dated references, only the edition cited applies. For undated references, the latest edition of the referenced document (including any amendments) applies.

GB/T 37526, *Assessment method for solar energy resource*

GB 50797, *Code for design of photovoltaic power station*

GB/T 6495.3, *Photovoltaic devices—Part 3: Measurement principles for terrestrial photovoltaic (PV) solar devices with reference spectral irradiance data*

*Land use control of photovoltaic power project* (GTZG [2015] No. 11)

IEC 61724-1, *Photovoltaic system performance—Part 1: Monitoring*

## 3 Terms and definitions

For the purposes of this document, the following terms and definitions apply.

### 3.1 standard test conditions

test conditions where the cell temperature is 25 °C, the solar irradiance is 1000 $W/m^2$ and the air mass is AM1.5

NOTE The spectral irradiance distribution meets the requirements of GB/T 6495.3.

### 3.2 nominal power of the rear side of a bifacial PV module

peak power of the rear side of the bifacial PV module measured under standard test conditions, expressed in Wp

### 3.3 installation capacity of PV modules

sum of nominal powers of PV modules installed in the photovoltaic power generation system for mono-facial PV modules, expressed in Wp

sum of nominal powers of the front side of bifacial PV modules, referred to as the front installation capacity, expressed in Wp; sum of nominal powers of the rear side of bifacial PV modules, referred to as the rear installation capacity, expressed in Wp. By the installation capacity of a photovoltaic power generation system consisting of bifacial modules, it is meant the front installation capacity

### 3.4 rated capacity

sum of rated active powers of inverters installed in the photovoltaic power generation system, expressed in W

### 3.5 PV power to inverter power ratio

ratio of the installation capacity of the photovoltaic power generation system to its rated capacity

### 3.6 mismatch losses

sum of the mismatch power losses in the series-connected PV modules, the parallel-connected PV module series and the parallel-connected combiner boxes

### 3.7 DC cable losses

sum of the power losses of all DC cables from PV module output terminals to inverter input terminals

### 3.8 AC cable losses

sum of the power losses of all AC cables from the inverter output terminals to the step-up transformer and from the step-up transformer to the point of coupling for the grid-connected photovoltaic power generation system

### 3.9 system failure and maintenance losses

electric energy loss resulting from failure, maintenance and malmanagement of modules, inverters, step-up transformers and other equipment of the photovoltaic power generation system

### 3.10 system external losses

electric energy loss of the photovoltaic power generation system resulting from external factors such as power grid and natural disasters

### 3.11 electric energy tariff point

metering point at the boundary between the photovoltaic power generation system and power grid or as specified in the contract

### 3.12 levelized cost of energy

ratio of all discounted costs of the photovoltaic power generation system to all discounted available on-grid energy within the assessment period, expressed in CNY/kWh

### 3.13 system performance

system of indicators used to comprehensively assess the techno-economy of a photovoltaic power generation system, including the annual equivalent full-load operation hours, energy efficiency ratio (EER), levelized cost of energy (LCOE), performance per unit area, external benefits, etc.

# 4 Performance parameters

## 4.1 Efficiency of PV module

The efficiency of PV module shall be the ratio of module power measured under standard test conditions to the product of module area and radiation intensity, and the calculation formula is as follows:

$$\eta_{\mathrm{m}} = \frac{P_{\mathrm{m}}}{S_{\mathrm{m}} \times G_0} \tag{1}$$

where

$\eta_{\mathrm{m}}$ is the efficiency of the module (%);

$P_{\mathrm{m}}$ is the power of the module (Wp);

$S_{\mathrm{m}}$ is the area of the module ($\mathrm{m}^2$);

$G_0$ is the radiation intensity under standard test conditions, a constant equal to 1000 W/m$^2$.

### 4.2 Bifaciality factor of bifacial PV module

The bifaciality factor of bifacial PV module shall be the ratio of the nominal power of the rear side of bifacial PV module to the nominal power of the front side of bifacial PV module, and the calculation formula is as follows:

$$\eta_{\text{BiFi}} = \frac{P_{\text{rear stc}}}{P_{\text{front stc}}} \times 100\ \% \tag{2}$$

where

$\eta_{\text{BiFi}}$ is the bifaciality factor of bifacial PV module (%);

$P_{\text{rear stc}}$ is the nominal power of the rear side of bifacial PV module (Wp);

$P_{\text{front stc}}$ is the nominal power of the front side of bifacial PV module (Wp).

### 4.3 Efficiency of inverter

The efficiency of inverter includes the maximum conversion efficiency and weighted average conversion efficiency. The maximum efficiency of the inverter is given in the certification issued by a specialized testing body. The weighted average gross efficiency of inverter shall be calculated by the following formulas:

$$\eta = \sum_{i=5\ \%}^{100\ \%} a_{\text{CHN}} \times \eta_{\text{conv},\ i} \times \eta_{\text{MPPTstat},\ i} \tag{3}$$

$$\eta_{\text{CHN}} = \frac{1}{N}\sum_{1}^{N}\eta \tag{4}$$

where

$\eta$ is the weighted efficiency of constant voltage;

$\eta_{\text{CHN}}$ is the weighted average gross efficiency;

$a_{\text{CHN}}$ is the weighting coefficient;

$\eta_{\text{conv},\ i}$ is the conversion efficiency of inverter measured under specified MPPT voltage;

$\eta_{\text{MPPTstat},\ i}$ is the static MPPT efficiency of inverter measured under specified MPPT voltage;

$N$ is the number of sampled data, $N$=5, i.e. five MPPT voltage values.

NOTE 1 $i$ is one of the numerals from 1 to 7, corresponding to loading points 5 %, 10 %, 20 %, 30 %, 50 %, 75 % and 100 %, respectively.

NOTE 2 The five MPPT voltages are $U_{\text{MPPMax}}$, $U_{\text{MPPMin}}+0.7\Delta U$, $U_{\text{MPPMin}}+0.5\Delta U$, $U_{\text{MPPMin}}+0.3\Delta U$ and $U_{\text{MPPMin}}$, where $\Delta U = U_{\text{MPPMax}} - U_{\text{MPPMin}}$.

The weighting coefficients of weighted efficiency of grid-connected inverters in China's solar energy resource areas are shown in Table 1.

**Table 1 Weighting coefficients of weighted efficiency of grid-connected inverters in China's solar energy resource areas**

| Loading point | 5 % | 10 % | 20 % | 30 % | 50 % | 75 % | 100 % |
|---|---|---|---|---|---|---|---|
| Weighting coefficient | $a_{\text{CHN-1}}$ | $a_{\text{CHN-2}}$ | $a_{\text{CHN-3}}$ | $a_{\text{CHN-4}}$ | $a_{\text{CHN-5}}$ | $a_{\text{CHN-6}}$ | $a_{\text{CHN-7}}$ |
| | 0.02 | 0.04 | 0.07 | 0.15 | 0.29 | 0.33 | 0.10 |

When calculating the weighted efficiency in a specific resource area, the efficiency may be calculated by Formula (3) using the weighting coefficient for the given resource area. The weighting coefficients in different solar energy resource areas are shown in Table 2.

**Table 2 Weighting coefficients in different solar energy resource areas**

| Loading point | 5 % | 10 % | 20 % | 30 % | 50 % | 75 % | 100 % |
|---|---|---|---|---|---|---|---|
| Weighting coefficient ($G \geq 6300$) | $a_{\text{CHN-1}}$ | $a_{\text{CHN-2}}$ | $a_{\text{CHN-3}}$ | $a_{\text{CHN-4}}$ | $a_{\text{CHN-5}}$ | $a_{\text{CHN-6}}$ | $a_{\text{CHN-7}}$ |
| | 0.01 | 0.02 | 0.05 | 0.12 | 0.26 | 0.35 | 0.19 |
| Weighting coefficient ($6300 > G \geq 5040$) | $a_{\text{CHN-1}}$ | $a_{\text{CHN-2}}$ | $a_{\text{CHN-3}}$ | $a_{\text{CHN-4}}$ | $a_{\text{CHN-5}}$ | $a_{\text{CHN-6}}$ | $a_{\text{CHN-7}}$ |
| | 0.02 | 0.04 | 0.07 | 0.16 | 0.29 | 0.32 | 0.10 |
| Weighting coefficient ($5040 > G \geq 3780$) | $a_{\text{CHN-1}}$ | $a_{\text{CHN-2}}$ | $a_{\text{CHN-3}}$ | $a_{\text{CHN-4}}$ | $a_{\text{CHN-5}}$ | $a_{\text{CHN-6}}$ | $a_{\text{CHN-7}}$ |
| | 0.03 | 0.04 | 0.08 | 0.17 | 0.31 | 0.31 | 0.06 |
| Weighting coefficient ($G < 3780$) | $a_{\text{CHN-1}}$ | $a_{\text{CHN-2}}$ | $a_{\text{CHN-3}}$ | $a_{\text{CHN-4}}$ | $a_{\text{CHN-5}}$ | $a_{\text{CHN-6}}$ | $a_{\text{CHN-7}}$ |
| | 0.05 | 0.08 | 0.12 | 0.18 | 0.25 | 0.30 | 0.02 |
| NOTE $G$ represents the annual global irradiation, and adopts the average annual value (generally over 30 years) [MJ/ ($m^2 \cdot a$)]. | | | | | | | |

## 4.4 PV power to inverter power ratio

The PV power to inverter power ratio shall be calculated in accordance with the following formula:

$$R = \frac{P_{\text{dc}}}{P_{\text{ac}}} \tag{5}$$

where

$R$ is the PV power to inverter power ratio;

$P_{\text{dc}}$ is the installation capacity of the system (Wp);

$P_{\text{ac}}$ is the rated capacity of the system (W).

See Annex A for the flow chart of PV power to inverter power ratio optimization calculation of photovoltaic power generation system, and Annex B for the case study of PV power to inverter power ratio optimization calculation of photovoltaic power generation system in typical areas.

## 4.5 DC power loss

The DC power loss of the system shall be the ratio of the sum of all mismatch losses and DC cable losses from module output to inverter input, to the sum of output powers of all modules, expressed as percentage.

## 4.6 AC power loss

For grid-connected photovoltaic power generation system, the AC power loss of the system shall be the ratio of the sum of all AC cable losses from the inverter output to the point of coupling and the power loss of step-up transformer, to the sum of all inverter output powers, expressed as a percentage.

## 4.7 Selection of solar energy resource data for photovoltaic power generation system

The calculation value of solar energy resources in photovoltaic power generation system should be prioritized based on the measured data of site solar radiation observation station for at least a full year. When the measured data of the site solar radiation observation station cannot be obtained, the calculation may be conducted on the basis of long-term observation data from representative meteorological stations or the calculated long-term radiation data, statistical calculation data based on satellite remote sensing data, or physical inversion data. The measurement and processing of radiation data shall be in accordance with IEC 61724-1. See Table 3 for the priority level of solar energy resource data selection for photovoltaic power generation system.

**Table 3 Priority level of solar energy resource data selection for photovoltaic power generation system**

| Data source | Priority level of data selection |
|---|---|
| Site observation station for solar radiation | A |
| Representative meteorological station | B |
| Satellite remote sensing data | C |

### 4.8 Calculated annual on-grid energy of photovoltaic power generation system

The calculated annual on-grid energy of photovoltaic power generation system shall comply with GB 50797.

### 4.9 Actual on-grid energy of photovoltaic power generation system

The actual on-grid energy shall be the measured value at the electric energy tariff point of photovoltaic power generation system within the assessment period.

### 4.10 Integrated auxiliary power ratio

The integrated auxiliary power ratio shall be the ratio of the integrated auxiliary energy consumption to the total energy at the inverter outlet within the assessment period, expressed as a percentage.

$$R_{\mathrm{s}} = \frac{E_{\mathrm{q}} - E_{\mathrm{p}} + E_{\mathrm{d}}}{E_{\mathrm{q}}} \times 100\ \% \qquad (6)$$

where

$R_{\mathrm{s}}$ is the integrated auxiliary power ratio (%);

$E_{\mathrm{q}}$ is the total energy at inverter outlet (kWh);

$E_{\mathrm{p}}$ is the on-grid energy (kWh);

$E_{\mathrm{d}}$ is the off-grid energy (kWh).

### 4.11 Total land occupation area for photovoltaic power generation system

The total land occupation area includes the actual land area for PV array (covering the area for multi-purpose items in the array), substation and operations center, collection lines and on-site roads.

### 4.12 Installation capacity of PV modules per unit area

The installation capacity of PV modules per unit area shall be the ratio of the installation capacity of photovoltaic power generation system to the total land occupation area of the system.

### 4.13 On-grid energy per unit area

The on-grid energy per unit area shall be the ratio of the total on-grid energy of photovoltaic power generation system within the assessment period to the total land occupation area of the system.

NOTE The assessment period should be taken as 25 years.

### 4.14 Power generation revenue per unit area

The power generation revenue per unit area shall be the ratio of the present value of total power generation revenue of photovoltaic power generation system over the assessment period to the total land occupation area of the system.

### 4.15 Economic revenue per unit area

The economic revenue per unit area shall be the ratio of the present value of total economic revenue of photovoltaic power generation system over the assessment period to the total land occupation area of the system.

NOTE The discount rate may be taken as 5 %, 8 % or the interest rate of long-term treasury bonds.

### 4.16 External benefits

The external benefits shall be the estimated value of emission reduction and eco-environmental benefits of photovoltaic power generation system within the assessment period. The external benefits include economic benefits and social benefits, etc., which may be estimated with reference to the market prices of relevant international or domestic carbon emission trading programs.

## 5 Performance indicators and assessment methods

### 5.1 Performance assessment content

The performance assessment is the process of calculating the performance indicators of photovoltaic power generation system by using performance parameters and comprehensively evaluating the performance of photovoltaic power generation system. The performance indicators include the generating capacity, EER, cost of energy, performance per unit area, external benefits, etc.

### 5.2 Performance assessment methods

**5.2.1** Annual equivalent full-load operation hours

The annual equivalent full-load operation hours is used to measure the generating capacity on the AC side of the photovoltaic power generation system, and calculated by using the accumulated on-grid energy, rated capacity, etc. within the assessment period.

The annual equivalent full-load operation hours of the system shall be calculated by the following formula:

$$A = \frac{\sum_{1}^{n} E_{\mathrm{t}}}{Y \times P_{\mathrm{ac}}} \tag{7}$$

where

$A$ is the annual equivalent full-load operation hours of the system (h);

$n$ is the number of hours over the performance assessment period of the system;

$E_{\mathrm{t}}$ is the hourly energy output within the assessment period of the system (kWh), $t = 1, 2, \ldots, n$;

$Y$ is the number of years over the performance assessment period of the system;

$P_{ac}$ is the rated capacity of the system (kW).

**5.2.2** Energy efficiency ratio (EER) of system

The EER of the system refers to the ratio of the on-grid energy to the theoretical energy output, which is used to measure the generating efficiency of the photovoltaic power generation system. When the bifacial PV modules are used, the theoretical energy output shall be the sum of the theoretical energy output of the front and rear sides of the PV array.

The EER of mono-facial PV module system shall be calculated by the following formula:

$$PR = \frac{\sum_{1}^{n} E_{p,i}}{\left(\sum_{1}^{n} H_i \times P_{dc}\right) / G_0} \tag{8}$$

where

$PR$ is the EER of the system;

$n$ is the number of hours over the performance assessment period of the system;

$E_{p,i}$ is the on-grid energy of the system over the assessment period (kWh), $i = 1, 2, ..., n$;

$H_i$ is the global irradiation on the front side of the PV array over the assessment period (kWh/m$^2$), $i = 1, 2, ..., n$;

$P_{dc}$ is the installation capacity of photovoltaic power generation system (kWp);

$G_0$ is the solar radiation intensity under standard conditions, a constant equal to 1 kW/m$^2$.

The EER of a bifacial PV module system shall be calculated by the following formula:

$$PR = \frac{\sum_{1}^{n} E_{p,i}}{\left(\sum_{1}^{n} H_{i,f} \times P_{dc,f} + \sum_{1}^{n} H_{i,r} \times P_{dc,r}\right) / G_0} \tag{9}$$

where

$PR$ is the EER of the system;

$n$ is the number of hours within performance assessment period of the system;

$E_{p,i}$ is the on-grid energy of the system over the assessment period (kWh), $i = 1, 2, ..., n$;

$H_{i,f}$ is the global irradiation on the front side of the PV array over the assessment period (kWh/m$^2$), $i = 1, 2, ..., n$;

$H_{i,r}$ is the global irradiation on the rear side of the PV array over the assessment period (kWh/m$^2$), $i = 1, 2, ..., n$;

$P_{dc,f}$ is the front installation capacity on the DC side of the photovoltaic power generation system, i.e. the sum of the nominal power of the front side of the modules (kWp);

$P_{dc,r}$ is the installation capacity on the rear side on the DC side of the system, which can be calculated by multiplying the sum of nominal power of the front side by the bifaciality factor (BF) (kWp);

$G_0$ is the solar radiation intensity under standard conditions, a constant equal to 1 kW/m$^2$.

When evaluating energy efficiency, the standard EER of the system under standard conditions may be used to analyze the photovoltaic power generation system, and the standard EER of the system shall be calculated by the following formula:

$$PR_{\text{stc}} = \frac{\sum_{1}^{n} E_{\text{p},i}}{\left(\sum_{1}^{n} H_{i,\text{f}} \times P_{\text{dc,f}} \times K_{\text{t}} + \sum_{1}^{n} H_{i,\text{r}} \times P_{\text{dc,r}} \times K_{\text{t}}\right) / G_0} \tag{10}$$

where

$PR_{\text{stc}}$ is the standard energy efficiency indicator (EEI) ratio of the system;

$K_{\text{t}}$ is the temperature correction coefficient over the assessment period (%).

The calculation formula of the temperature correction coefficient is as follows:

$$K_{\text{t}} = 1 + \gamma\,(T_{\text{mod,n}} - 25) \tag{11}$$

where

$\gamma$ is the relative temperature coefficient of module power (%);

$T_{\text{mod,n}}$ is the operating temperature of modules over the assessment period (°C).

**5.2.3** Levelized cost of energy

The LCOE is calculated by the following formula:

$$LCOE = \left[ I_0 - \sum_{n=1}^{N} \frac{I_{\text{t}}}{(1+i)^n} - \frac{V_{\text{R}}}{(1+i^N)} + \sum_{n=1}^{N} \frac{M_n}{(1+i)^n} \right] / \sum_{n=1}^{N} \frac{Y_n}{(1+i)^n} \tag{12}$$

where

$LCOE$ is the levelized cost of energy (CNY/kWh);

$i$ is the discount rate (%);

$n$ is the number of operation years of the system, $n$ = 1, 2 ,..., $N$;

$N$ is the assessment period of the system (a);

$I_0$ is the static initial investment of the system (CNY);

$I_{\text{t}}$ is the project VAT deduction (CNY);

$V_{\text{R}}$ is the residual value of the system (CNY);

$M_n$ is the operating cost in the $n$th year (including maintenance, insurance, materials, labor wages, auxiliary service charges, etc., interest not included) (CNY);

$Y_n$ is the annual on-grid energy (kWh).

**5.2.4** Performance per unit area

The indicators for performance per unit area mainly include the installation capacity per unit area, energy output per unit area, power generation revenue per unit area and economic revenue per unit area. The calculation of the indicators shall meet the following requirements:

**a)** The installation capacity per unit area shall be calculated by the following formula:

$$PS_{\mathrm{p}} = \frac{P_{\mathrm{dc}}}{S} \tag{13}$$

where

$PS_{\mathrm{p}}$ is the installation capacity per unit area (Wp/$m^2$);

$P_{\mathrm{dc}}$ is the installation capacity of the system (Wp);

$S$ is the total land occupation area of the system, including the land area used for PV array, substation and operations center, power collector lines and on-site roads ($m^2$).

**b)** The energy output per unit area shall be calculated by the following formula:

$$PS_{\mathrm{s}} = \frac{\sum_{1}^{n} E_{\mathrm{p}}}{S} \tag{14}$$

where

$PS_{\mathrm{s}}$ is the energy output per unit area (kWh/$m^2$);

$\sum_{1}^{n} E_{\mathrm{p}}$ is the on-grid energy of the system in assessment period (kWh);

$S$ is the total land occupation area of the system, including the land area used for PV array, substation and operations center, power collector lines and on-site roads ($m^2$).

**c)** The power generation revenue per unit area shall be calculated by the following formula:

$$PS_{\mathrm{e}} = \left[\sum_{n=1}^{N} \frac{PG_{\mathrm{d},n}}{(1+i)^n}\right] / S \tag{15}$$

where

$PS_{\mathrm{e}}$ is the power generation revenue per unit area (CNY/$m^2$);

$PG_{\mathrm{d},n}$ is the revenue from electricity sale in the $n$th year (CNY);

$i$ is the discount rate (%);

$S$ is the total land occupation area, including the land area used for PV array, substation and operations center, collector lines and on-site roads ($m^2$).

**d)** The economic revenue per unit area shall be calculated by the following formula:

$$PS_{\mathrm{c}} = \left[\sum_{n=1}^{N} \frac{PG_{\mathrm{d},n}}{(1+i)^n} + \sum_{n=1}^{N} \frac{PG_{\mathrm{z},n}}{(1+i)^n}\right] / S \tag{16}$$

where

$PS_{\mathrm{c}}$ is the economic revenue per unit area (CNY/$m^2$);

$PG_{\mathrm{d},n}$ is the revenue from electricity sale in the $n$th year (CNY);

$PG_{\mathrm{z},n}$ is the revenue from comprehensive utilization of the system in the $n$th year (CNY);

$i$ is the discount rate (%);

$S$ is the total land occupation area, including the actual land area used for PV array, substation and operations center, power collector lines and on-site roads ($m^2$).

**5.2.5** External benefit per kilowatt-hour

The external benefit per kilowatt-hour is calculated by the following formula:

$$P_{ex} = \frac{\sum_{1}^{n} PG_{ex}}{\sum_{1}^{n} E_{p}} \tag{17}$$

where

$P_{ex}$ is the external benefit per kilowatt-hour (CNY/kWh);

$PG_{ex}$ is the external benefit of the system over the assessment period (CNY);

$\sum_{1}^{n} E_{p}$ is the on-grid energy over the assessment period (kWh).

## 5.3 Performance assessment process

**5.3.1** General

The performance assessment process should be divided into such steps as determining assessment objectives, selecting assessment indicators, collecting data, carrying out assessment, giving conclusions, making suggestions on improvement, and preparing the assessment report.

**5.3.2** Data collection

The data over an operation period of at least one year should be collected as the basic data for performance assessment of the photovoltaic power generation system. The following data needs to be collected:

**a)** System rated capacity.

**b)** System installation capacity (front and rear installation capacities need to be collected if bifacial PV modules are used).

**c)** Array surface irradiation (front and rear irradiations need to be collected if bifacial PV modules are used).

**d)** Energy output on inverter AC side.

**e)** On-grid energy.

**f)** System failure and maintenance losses and system external losses.

**g)** Temperature of rear panel of modules.

**h)** Temperature coefficient of modules.

**i)** Land occupation area.

**j)** Total investment.

**k)** Financial cost over the assessment period.

**l)** Operating cost and other costs.

**m)** System energy output required by the power market or the power grid and the economic profit and loss.

**n)** System revenues from electricity sale and comprehensive utilization.

**o)** System external benefits.

**5.3.3** Performance assessment

Based on the basic data, calculate the relevant parameters and indicators in accordance with the performance analysis method. Through the study of the calculation results, draw the assessment conclusions on the current performance of photovoltaic power generation system, put forward suggestions on improvement, and complete the assessment report. The assessment report should include project profile, basic data, data analysis, calculation of performance parameters and indicators, assessment conclusions, suggestions on improvement, etc.

# Annex A
(normative)
# Optimization of PV power to inverter power ratio of photovoltaic power generation system

The PV power to inverter power ratio of photovoltaic power generation system should be calculated and optimized through techno-economic comparison, comprehensively considering the geographical position of the project, topographic conditions, solar energy resources, module selection, installation type, layout pattern, inverter performance, construction cost, losses from photovoltaic array to inverter or point of coupling, power grid demand, etc.. The optimization analysis of PV power to inverter power ratio should use the trial and error method, and the optimal PV power to inverter power ratio should be obtained through multiple calculations by taking PV power to inverter power ratio in ascending order. See Figure A for the flowchart of optimization of PV power to inverter power ratio of photovoltaic power generation system.

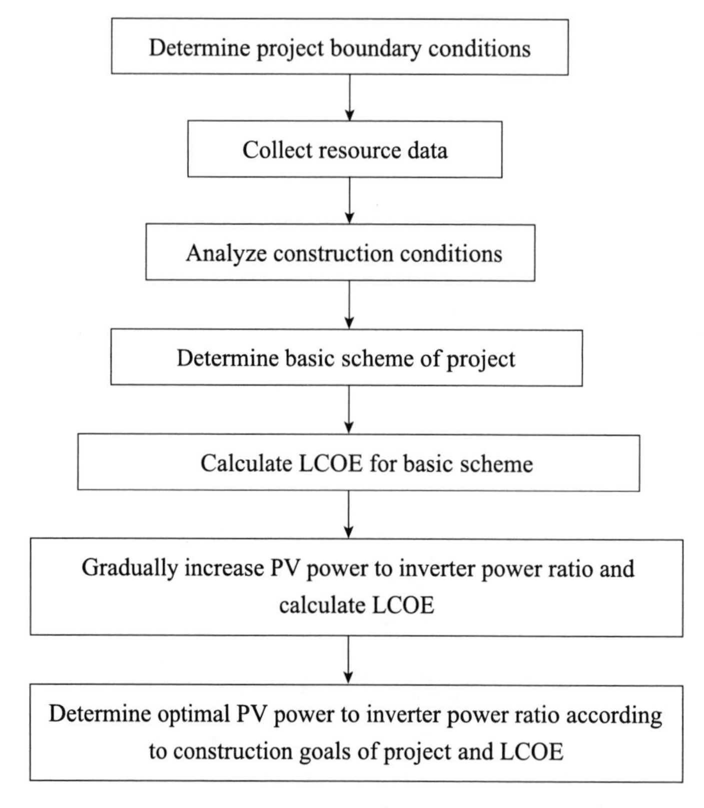


**Figure A Flowchart of optimization of PV power to inverter power ratio of photovoltaic power generation system**

# Annex B
(informative)
# Example for optimization of PV power to inverter power ratio of photovoltaic power generation system in typical regions

## B.1 Selection of typical region

Solar energy resources are classified into four grades according to the global horizontal radiation (*GHR*) grade table given in GB/T 37526, as shown in Table B.1. Select three representative areas with the highest, lowest and median values from each grade, conduct the analysis of optimal PV power to inverter power ratio, and divide the range of 900 kWh/m$^2$ to 2000 kWh/m$^2$ into 12 intervals with an increment of 100 kWh/m$^2$ and select respective representative areas for analysis, as shown in Table B.2. By comparing the radiation data from each representative meteorological station, select the meteorological data of recent 10 years as the actual representative radiation data, and discretize the data of representative year into hourly radiation data to calculate the energy output.

**Table B.1 Grades of annual global horizontal radiation (*GHR*)**

| Grade | Grade threshold /(MJ/m$^2$) | Grade threshold /(kWh/m$^2$) | Symbol |
|---|---|---|---|
| Extremely rich | $GHR \geq 6300$ | $GHR \geq 1750$ | A |
| Very rich | $5040 \leq GHR < 6300$ | $1400 \leq GHR < 1750$ | B |
| Rich | $3780 \leq GHR < 5040$ | $1050 \leq GHR < 1400$ | C |
| Fair | $GHR < 3780$ | $GHR < 1050$ | D |

**Table B.2 Statistics of annual *GHR* of representative locations from different sources**

| S/N | Global radiance /(kWh/m$^2$) | Region | Meteorological station (1988 - 2007) /(kWh/m$^2$) | Meteorological station (1998 - 2007) /(kWh/m$^2$) |
|---|---|---|---|---|
| 1 | 900 | Chongqing | 868.9 | 875.4 |
| 2 | 1000 | Mianyang | 969 | 979 |
| 3 | 1100 | Changsha | 1075.1 | 1048.1 |
| 4 | 1200 | Hefei | 1209.6 | 1232.6 |
| 5 | 1300 | Ji'nan | 1298.8 | 1300.7 |
| 6 | 1400 | Changchun | 1383.4 | 1366.1 |
| 7 | 1500 | Kunming | 1521.8 | 1533.7 |
| 8 | 1600 | Yinchuan | 1621.2 | 1588.2 |
| 9 | 1700 | Minqin | 1730.8 | 1736.4 |
| 10 | 1800 | Dunhuang | 1781.5 | 1781.9 |
| 11 | 1900 | Golmud | 1928.5 | 1897 |
| 12 | 2000 | Lhasa | 1980 | 2056.3 |

## B.2 Calculation boundary conditions of typical cases

For the typical case, the following boundary conditions apply:

**a)** Take the 100 MW photovoltaic power generation system as an example, keep the rated (AC) capacity of 100 MW unchanged, increase the installation capacity by an interval of 0.1 within the PV power to inverter power ratio of 1.1 to 1.9 for analyzing the cost per kWh.

**b)** For energy output calculation, select the mono-crystalline 420 Wp PV modules (mono-facial or bifacial), 3125 kW centralized inverters or 175 kW string inverters. The ground reflectivity is 30 %, and the module ground clearance is 1.0 m.

**c)** The power at point of coupling of photovoltaic power generation system is limited at 100 MW, and other limitations such as surplus PV energy output rate and guaranteed utilization hours of photovoltaic power generation system are not considered.

**d)** Keep the cost on AC side of photovoltaic system unchanged for different PV power to inverter power ratios, and calculate the cost on DC side according to the actual bill of quantities.

**e)** The land occupation of PV sub-array layout complies with Document GTZG [2015] No. 11, *Land use control of photovoltaic power project*, issued by the Ministry of Land and Resources in 2015.

**f)** Financial parameter values for cost per kWh analysis: The depreciation period of the project is 15 years (with a residual value of 5 %), and the calculation base is the original value of fixed assets; the repair rate is 0.2 %; the per capita annual salary of employees is 100000 CNY; employee welfare, labor insurance and housing fund is 60 % of total wages; the insurance premium rate is 0.25 % of the value of fixed assets; material cost quota is 10 CNY/kW; the quota of other fees is 25 CNY/kW; the staffing is 5 persons. The auxiliary service cost is not considered.

**g)** The PV modules are arranged on a sun slope of not greater than 25°.

**h)** The influence of the differences in irradiation and temperature between the actual site and the meteorological station is not considered.

**i)** The PV power to inverter power ratio in the scheme with the lowest cost per kWh is considered optimal, which is corrected considering the actual irradiation at the representative sites to obtain the ratio for the typical region.

## B.3 Calculation process of typical cases

Take Hefei (the global irradiation is 1232.6 kWh/m$^2$) as the typical region in central China, and adopt the scheme of fixed type + mono-facial modules + centralized inverters to calculate the LCOEs for different PV power to inverter power ratios according to the proposed boundary conditions. See Table B.3.

**Table B.3 LCOE calculation results for different PV power to inverter power ratios**

| Item | Unit | Ratio 1.4 | Ratio 1.5 | Ratio 1.6 | Ratio 1.7 | Ratio 1.8 |
|---|---|---|---|---|---|---|
| Rated capacity | MW | 100.00 | 100.00 | 100.00 | 100.00 | 100.00 |
| Installed capacity | MWp | 138.38 | 150.96 | 159.34 | 171.92 | 180.31 |
| Static project investment | $10^3$ CNY | 558855.6 | 605710.8 | 637186.0 | 684060.8 | 715540.2 |

**Table B.3** *(continued)*

| Item | Unit | Ratio 1.4 | Ratio 1.5 | Ratio 1.6 | Ratio 1.7 | Ratio 1.8 |
|---|---|---|---|---|---|---|
| Static investment per kilowatt | CNY/kWp | 4038.61 | 4012.44 | 3998.79 | 3978.84 | 3968.37 |
| Annual equivalent full-load operation hours (capacity DC side) | h | 1016 | 1013 | 1008 | 998 | 989 |
| Annual equivalent full-load operation hours (capacity AC side) | h | 1405 | 1529 | 1607 | 1716 | 1784 |
| LCOE | CNY/kWh | 0.4962 | 0.4908 | 0.4893 | 0.4896 | 0.4913 |

It can be seen from Table B.3 that LCOE is the lowest when the PV power to inverter power ratio is 1.6, so 1.6 is selected as the optimal PV power to inverter power ratio.

## B.4 Calculation results for typical region

Under the boundary conditions specified in Table B.2, the PV power to inverter power ratios of typical projects in typical region with different irradiations under different operation modes are calculated, as shown in Tables B.4 and B.5 for reference.

**Table B.4 Calculation results of PV power to inverter power ratio of mono-facial modules in typical region**

| S/N | *GHR* /(kWh/m$^2$) | Horizontal | Fixed | Horizontal single-axis tracking | Oblique single-axis tracking |
|---|---|---|---|---|---|
| 1 | 1000 | 1.7 - 1.8 | 1.7 - 1.8 | 1.6 - 1.7 | 1.5 - 1.6 |
| 2 | 1200 | 1.7 | 1.6 - 1.7 | 1.6 | 1.5 |
| 3 | 1400 | 1.6 | 1.5 - 1.6 | 1.5 | 1.4 |
| 4 | 1600 | 1.4 | 1.4 | 1.4 | 1.3 |
| 5 | 1800 | 1.3 - 1.4 | 1.3 | 1.3 - 1.4 | 1.2 - 1.3 |
| 6 | 2000 | 1.2 | 1.1 - 1.2 | 1.1 - 1.2 | 1.0 - 1.1 |

**Table B.5 Calculation results of PV power to inverter power ratios of bifacial modules in typical region**

| S/N | *GHR* /(kWh/m$^2$) | Fixed | Horizontal single-axis tracking | Oblique single-axis tracking |
|---|---|---|---|---|
| 1 | 1000 | 1.6 - 1.7 | 1.5 - 1.6 | 1.5 |
| 2 | 1200 | 1.6 | 1.5 - 1.6 | 1.4 |
| 3 | 1400 | 1.5 | 1.4 - 1.5 | 1.3 - 1.4 |
| 4 | 1600 | 1.3 | 1.3 - 1.4 | 1.2 - 1.3 |
| 5 | 1800 | 1.2 - 1.3 | 1.3 | 1.2 |
| 6 | 2000 | 1.1 | 1.0 - 1.2 | 1.0 |

Such factors as the latitude of PV power project site, the difference of monthly irradiation, construction conditions (temperature, terrain and altitude, etc.), financial model and guaranteed utilization hours will affect the calculation results of optimal PV power to inverter power ratio. Taking Hegang (1274 kWh/m$^2$) as an example, the *GHR* is similar to that of Hefei (1232.6 kWh/

m$^2$); however, with the influence of such factors as latitude, difference of monthly irradiation and construction conditions, the calculated optimal PV power to inverter power ratio of Hegang is 1.5 when fixed mono-facial PV modules are used and 1.4 when fixed bifacial PV modules are used under the boundary conditions set in Section B.2. While carrying out engineering design, the calculations must be made according to the actual conditions of the project to select the appropriate PV power to inverter power ratio.